ANTIQUES AND THEIR VALUES

BRONZES

Compiled by
TONY CURTIS

D0815477

First Published   July   1976
Reprinted   Jan   1977
    ..   Oct   1977
Revised Edition June   1978

Exchange Rate $2 = £1

First Edition   ISBN 0-902921-40-1
Revised Edition ISBN 0-902921-71-1

INTRODUCTION

Congratulations! You now have in your hands an extremely valuable book. It is one of a series specially devised to aid the busy professional dealer in his everyday trading. It will also prove to be of great value to all collectors and those with goods to sell, for it is crammed with illustrations, brief descriptions and valuations of hundreds of antiques.

Every effort has been made to ensure that each specialised volume contains the widest possible variety of goods in its particular category though the greatest emphasis is placed on the middle bracket of trade goods rather than on those once - in - a - lifetime museum pieces whose values are of academic rather than practical interest to the vast majority of dealers and collectors.

This policy has been followed as a direct consequence of requests from dealers who sensibly realise that, no matter how comprehensive their knowledge, there is always a need for reliable, up-to-date reference works for identification and valuation purposes.

When using your Antiques and their Values to assess the worth of goods, please bear in mind that it would be impossible to place upon any item a precise value which would hold good under all circumstances. No antique has an exactly calculable value; its price is always the result of a compromise reached between buyer and seller, and questions of condition, local demand and the business acumen of the parties involved in a sale are all factors which affect the assessment of an object's 'worth' in terms of hard cash.

In the final analysis, however, such factors cancel out when large numbers of sales are taken into account by an experienced valuer, and it is possible to arrive at a surprisingly accurate assessment of current values of antiques; an assessment which may be taken confidently to be a fair indication of the worth of an object and which provides a reliable basis for negotiation.

Throughout this book, objects are grouped under category headings and, to expedite reference, they progress in price order within their own categories. Where the description states 'one of a pair' the value given is that for the pair sold as such.

Printed by Apollo Press, Dominion Way, Worthing, Sussex, England.
Bound by Newdigate Press, Vincent Lane, Dorking, Surrey, England.

CONTENTS

One of a pair of Victorian, bronzed spelter lions, 15in. long. $200 £100

19th century bronze of a charging oryx. $300 £150

A fine 19th century bronze of a buffalo, initialled. $360 £180

Cambodian gilded bronze antelope, reclining, circa 1880. $390 £195

Late 19th century bronze lion, signed Barye. $440 £220

One of a pair of late 19th century Chinese cloisonne bears. $460 £230

One of a pair of bronze rabbits by A. L. Barye, 2¾in. long.
$480 £240

Late 19th century bronze panther, signed Van de Kemp, 4½in. high.
$480 £240

A bronze cow with raised head, signed I. Bonheur, circa 1870, 6¼in. high. $720 £360

Late 19th century Japanese bronze figure of a tiger, 28in. long, signed.
$780 £390

Bronze figure of a cat, signed Jean Earril, 19in. high.
$1,100 £550

Large Egyptian bronze statue of a cat. $52,000 £26,000

Bronze statue of a bison by J. Haehnek, 12in. long. $300 £150

Late 19th century bronze bull, 8in. wide, signed I. Bonheur.
$500 £250

Bronze figure of a bull by P.J. Mene, with a rich gold patina. $720 £360

Japanese bronze figure of a running bull, pad mark with incised character marks, 29in. long, on wooden stand. $880 £440

One of a pair of cloisonne bulls, in turquoise, blue and red, 9in. long, 6½in. high. $1,860 £930

One of a fine pair of bronze bulls by Isadore Bonheur.
$3,300 £1,650

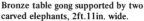

Late Victorian bronze elephant
paperweight, 3½in. high. $52 £26

Bronze table gong supported by two
carved elephants, 2ft.11in. wide.
$90 £45

Late 19th century Seiya bronze of
an elephant, Japanese, 14in. long.
$320 £160

A large 19th century bronze group of
Janesha, seated on the back of a three
headed elephant. $500 £250

A reproduction of a bronze Senegal
elephant by Barbedienne, cast from
the original mould. $900 £450

An original Senegal elephant by
Barye. $1,920 £960

19th century bronze of a stag with
two does. $290 £145

Good bronze stag by Falkirk, 24in.
high. $660 £330

Bronze group of three hounds attacking a stag, signed by P.J. Mene.
                           $1,080 £540

Late 19th century bronze group of a
stag standing guard over a drinking
doe. $2,600 £1,300

Mid 1st century B.C. Luristan bronze
of two stags, 10.9cm. high.
                        $17,400 £8,700

Celtic bronze deer of the 1st century
B.C., 4in. long.
                        $46,000 £23,000

Fine 18th century, green patinated bronze of a lion attacking a horse, mounted on a square stepped, rouge de fer marble base, 8in. high.
$250 £125

A 19th century Japanese bronze group of two tigers attacking a rhinoceros, 24in. wide. $350 £175

Bronze of two bears 'The Dentist' by Christopher Fratin, 4¾in. high.
$620 £310

Bronze figure of a 'Mare playing with a Dog' by P.J. Mene. $1,200 £600

Late 19th century bronze elephant attacked by tigers, 39cm. long.
$2,200 £1,100

An animalier bronze by Paul Edouard Delabrierre, 1802-1912, signed.
$2,500 £1,250

13

Art Nouveau gilt metal figure of a
girl dancer, 11in. high. $100 £50

An Art Nouveau coloured metal
figure of a girl and a pillar, on
onyx base, 11in. high. $150 £75

Late 19th century Russian model
of a lady. $180 £90

An Art Deco bronze, ivory and onyx
ash tray. $200 £100

Bronze and ivory Art Deco figure.
                    $200 £100

Bronze and ivory Art Deco female
figure. $220 £110

14

Spelter figure of a girl on an onyx and marble base, circa 1920, 1ft. 6in. long. $260 £130

One of a pair of Art Deco gilded metal figures of dancers, 14½in. high. $240 £120

An Art Deco bronze group by Etling. $300 £150

A French gilt bronze female on a white marble base, signed Cl. J. R. Colinet, 14½in. high. $360 £180

Art Nouveau bronze and ivory statuette of a dancer, 12in high. $380 £190

A bronze figure of a dancing cymbals player by Pilkington Jackson, signed and dated 1922, 2ft. 6in. high. $400 £200

A bronze dancing figure of Isadora
Duncan, 18in. high. $400 £200

An Art Nouveau bronze of a female
dancer poised with an ivory ball
balanced on her forearm, signed
Jaeger. $480 £240

Signed Art Nouveau bronze figure.
                          $480 £240

Art Deco bronze nude dancer
signed Bouraine, 23in. high.
                          $480 £240

Art Deco Bronze archer.
                  $540 £270

Silvered metal female figure signed
Fayral. $540 £270

Pair of French Art Nouveau gilt bronze Dutch figures with ivory faces and hands by 'La Monica', 33cm. high.
$550 £275

A finely modelled bronze figure depicting the actress Isadora Duncan in an Art Nouveau dress, signed on base 'Jaeger', 18½in. high.
$560 £280

Bronze and ivory figure of a young girl by Preiss, 8¼in high.
$560 £280

Art Deco group in silvered bronze.
$600 £300

Art Nouveau bronze of a female by Colinet.
$600 £300.

Bronze and ivory Art Deco figure.
$620 £310

17

Art Deco bronze and ivory group of
a seated female with two dogs. $640
£320

Chiparus gilt bronze and ivory figure
on green onyx base, 11in. high.
$640 £320

Bronze and ivory dancer by Cl. J.
R. Colinet, 14½in. high.
$760 £380

Art Nouveau bronze and ivory
'Dancing Girl' by Colinet. $800 £400

A finely modelled bronze figure,
signed on base in script, Jaeger ,
standing on dark brown marble
plinth, 18½in. high. $840 £420

Fine Art Nouveau painted bronze
and ivory figure of a dancer in thea-
trical costume. $900 £450

A stylish Art Deco bronze by Le Faguays. $900 £450

The Sun Worshipper, by Preiss, in bronze and ivory. $960 £480

Figure of a young girl with hoop, by Preiss. $960 £480

Preiss figure of a young boy with his hands in his pockets. $960 £480

An Art Nouveau bronze and ivory figure of a dancer by Demetre H. Chiparus, on an onyx base, 17in. high. $1,010 £505

A bronze group of a woman and a greyhound, by R. Rivoire, 1925. $1,080 £540

Painted metal Chiparus group of woman and goats. $1,100 £550

The Bathing Girl by Preiss, bronze and ivory on marble base, 1930's, 9¼in high. $1,100 £550

Bronze and ivory figure of a dancer by Philippe, 16½in high.
$1,300 £650

Bronzed metal and ivory group 'Toujours Des Amis' by Chiparus.
$1,300 £650

Art Deco bronze and ivory group of a horse with young girl attendant, on an onyx base, 1ft. 3in. wide.
$1,500 £750

Art Deco ivory and gilt bronze figure by Chiparus. $1,560 £780

Art Deco ivory and gilt bronze
figure by Chiparus. $1,560 £780

Art Deco, tinted bronze and ivory
group by Chiparus. $1,560 £780

A superb Art Deco female figure with
two dogs in bronze and ivory.
                              $1,680 £840

Preiss figure of a girl skating, 14in.
high. $1,680 £840

Bronze and ivory figure of a dancer
by D. H. Chiparus, 13in high.
                              $1,700 £850

'The Javelin Thrower' by F. Preiss,
12in. high. $1,800 £900

Ivory and gilt-bronze figure of a running girl, by F. Preiss.

$1,800 £900

Preiss figure of a standing girl in a casual dress.
$1,800 £900

A bronze and ivory figure of a dancer, by Chiparus, on a rouge marble base, 38.5cm. high. $1,870 £935

A fine Art Nouveau bronze and ivory figure of a ballet dancer, by Demetre H. Chiparus, on an onyx base, 19in. high.
$2,020 £1,010

Preiss bronze and ivory figure of a dancing girl, 15in. high.
$2,200 £1,100

Bronze and ivory figure by D. H. Chiparus, 11¾in high.
$2,200 £1,100

Bronze and ivory group by D. H. Chiparus, 11¼in high.
$2,300 £1,150

Art Deco bronze and ivory figure of an airwoman, by F. Preiss.
$2,400
£1,200

Bronze and ivory group 'The Bathers' by F. Preiss.
$2,400 £1,200

Bronze and ivory female figure by Chiparus.
$2,400 £1,200

Ivory and bronze figure 'Invocation' by Preiss.
$2,400 £1,200

'Con Brio', gilt bronze and ivory female figure by F. Preiss.
$2,400 £1,200

23

Bronze 'Flute Player' by F. Preiss, 17¾in. high. $2,420 £1,210

Untypical ivory and bronze figure by Franz Preiss. $2,500 £1,250

Art Deco bronze and ivory figure of a fan-dancer by Philippe, 18½in high. $2,800 £1,400

Ivory and bronze figure of a girl in 1920's costume, by F. Preiss, 14in. high. $2,860 £1,430

An ivory and bronze female figure by Chiparus. $2,860 £1,430

Bronze and ivory figure of the 'Torch Dancer' by Preiss, 15¾in. tall. $3,080 £1,540

Bronze and ivory figure of a dancer by D. H. Chiparus, 15½in high.
$3,200 £1,600

'The Balloon Girl' by F. Preiss, on an onyx base, 37cm. high.
$4,180 £2,090

Chiparus bronze and ivory dancer.
$4,200 £2,100

An exotic ivory and bronze figure of a dancer, on a shaped and coloured onyx base, 16in. high, signed Chiparus.
$4,400 £2,200

Gilt bronze figure of the dancer Louie Fuller by Raoul Larche, 18¼in. high. $7,200 £3,600

Bronze and ivory figure by Demetre Chiparus. $10,400 £5,200

Large bronze muffin man's hand-bell by Mears and Son. $70 £35

Oriental bronze table bell, on carved and pierced ironwood stand. $90 £45

A large bronze bell, inscribed John C. Wilson, Founder, Glasgow, 1860, 20½in. high, 10in. diam. $220 £110

Georgian bronze bell with a raised border that reads Arnold Rester,Fecit 1796. $260 £130

17th century bronze bell with cast inscription,Fudli I. Borchard. $360 £180

Japanese bronze bell of the Yayoi period, 62cm. high. $78,000 £39,000

Large cast hen on a nest circa 1845, 16in. long. $170 £85

A bronze pheasant, circa 1860.
                                    $170 £85

Bronze crane by August-Nicholas Cain, 4ins. high. $260 £130

A bronzed eagle with outstretched wings, perched on a branch, signed A. Thorburn, early 20th century, 4½in. long. $300 £150

One of a pair of bronze figures of cock and hen pheasants, 11¾in. high. $340 £170

One of a pair of Chinese cloisonne enamel quail. $360 £180

Pair of Oriental bronze cranes,
3ft. 3ins. high. $720 £360

Bronze fish eagle with shakudo beak
and crystal eyes, signed Mitani,
42.5cm. high. $850 £425

Japanese bronze figure of an eagle,
29in. wingspan, on a wooden base.
$860 £430

Japanese bronze eagle on tree
trunk with snake at base.
$880 £440

Weasel and cock pheasant bronze
group, inscribed 'J. Moigniez', 21in.
high. $900 £450

Pair of bronze partridges by E.
Pautrot, signed, 13½in. high.
$900 £450

Pair of Japanese metalware quail in copper, Shakudo and gilt, 12cm. high.
$960 £480

19th century bronze of 'Hens' by Andre Leonard and Garnier Freres.
$960 £480

A pair of Chinese cloisonne enamel cranes.
$960 £480

A large 19th century Chinese bronze crane, 7ft. high.
$960 £480

onze pheasant by A.A. Arson, 17in. gh.
$1,020 £510

One of a pair of Gyoko bronze mandarin ducks, late 19th century, 8¾in. high.
$1,200 £600

BENIN BRONZES

An unrecorded 16th century Benin cast bronze of a standing figure.
$26,000 £13,000

Nigerian Benin bronze of a Warrior on Horseback, 59.7cm. high.
$40,000 £20,00

A superb Benin bronze head, 10½in. high.
$46,000 £23,000

Benin bronze head of an Oba.
$64,000 £32,00

BOTTLES

A Victorian bronze bottle with figures of lions in relief, 11¾in. high. $40 £20

19th century bronze bottle, decorate with flowers, insects and serpents in high relief, signed on base W. Sherrif and dated 1895, 14½in. high. $110

Chinese bronze circular bowl with in-laid cloisonne banding, 7in. diam.
$60 £30

An Art Nouveau bronze bowl with flowers in relief, 24cm. long. $70 £35

Japanese bronze circular bowl with large rim, 8¾in.
$80 £40

A cloisonne enamel circular bowl, 7in. diam.
$90 £45

Kin Luong cloisonne enamel circular deep dish, 17¾in. diam.
$160 £80

A large lidded cloisonne bowl of circular shape, banded with cloud collars, 13½in. diameter.
$200 £100

31

BOWLS

One of a pair of turquoise enamel oval shaped bowls, 26cm. wide. $460 £230

Louis XVI ormolu mounted white marble brule parfum, 1ft. 5in. high. $1,900 £950

One of a pair of Louis XVI ormolu mounted celadon bowls, 10in. high. $6,800 £3,400

10th century Javanese bronze bowl, 20.9cm. high. $14,000 £7,000

BURMESE

A Burmese bronze figure of a goddess with traces of red gilt decoration, on a square base with four feet, 17in. high. $70 £35

19th century Burmese gilt bronze Buddha, 17in. high. $200 £10●

Small Victorian bronze bust of a
gentleman, 6in. high. $70 £35

19th century classical bronze bust,
7in. high. $100 £50

A bronze bust, circa 1840, (unsigned),
2in. high overall. $100 £50

One of a pair of 19th century
French bronze busts on marble
bases, 12in. high. $120 £60

th century bronze bust with Eastern
le head-dress. $120 £60

A small late 19th century bust of
Judith, 7in. high. $120 £60

33

BUSTS

A bronze statuary head of a lady with her hair tied in a knot, unsigned, 23in. high. $130 £65

A bronze statuary bust of a satyr, inscribed monogram, W.S., 12¾in. high. $160 £80

A German bronze bust of Kaiser Wilhelm II, 9in. high. $170 £85

Bronzed bust of Hitler, 11in. high, on a marble base. $170 £8⁅

Art Nouveau spelter bust, signed, circa 1900. $200 £100

Victorian bronze bust of Byron. $200 £100

French ivory bust of a girl, with ormolu drapery, signed A. Leonard, 24cm. high. $290 £145

A painted insurance figurehead, dated 1857. $420 £210

A large bronze of Kaiser Wilhelm l wearing full uniform and decorations, 17in. high. $440 £220

A well modelled heavy bronze figure of Goering, 15in. high. $460 £230

One of a pair of late 19th century busts of Moors, impressed Giesecke, 22in. high. $940 £470

Late 19th century coloured bronze bust of La Sibylle by E. Villanis, 28in. high. $1,000 £500

35

Louis XV ormolu portrait of
Madame de Pompadour, 1ft. high.
$1,600 £800

A bronze, by Rodin, 'La Pleureuse',
circa 1889, 11½in. high.
$5,600 £2,800

Silver bronze and parcel-gilt bust by
Alphonse Mucha. $15,000 £7,500

One of a set of twelve late 17th
century bronzes of Roman
Emperors, 29.2cm. high.
$20,000 £10,000

An important gilt bronze bust by
Alphonse Mucha, signed, 27½in.
high. $25,000 £12,500

Hellenistic or Roman bronze head of
a boy, 9¼in. high. $38,000 £19,000

A French ormolu candelabrum for three lights, on triangular shaped base, 19¾in. high. $170 £85

Fine pair of ormolu and bronze candelabra, circa 1815. $240 £120

Pair of 19th century bronze figure candelabra. $360 £180

Pair of 19th century bronze and ormolu candelabra. $380 £190

Pair of Regency bronze and ormolu three branch, four light candelabra. $730 £365

Pair of Adam Carrara marble, bronze and original fired-gilt candelabra, 15in. high, circa 1775. $740 £370

CANDELABRA

Pair of bronze and ormolu candelabra. $840 £420

Pair of French 'Egyptian' bronze and
ormolu,twin-branch candelabra.
                                  $850 £425

Pair of French Empire bronze and
ormolu candelabra, 23in. high.
                                  $960 £480

Pair of early 19th century bronze and
ormolu,seven branch candelabra, 25in.
high. $960 £480

A fine pair of Louis Philippe bronze
and ormolu candelabra. $1,140 £570

A pair of Louis XVI ormolu
mounted candelabra, 1ft.9½in.
high. $1,300 £650

38

A pair of ormolu mounted porphyry candelabra, 1ft.9½in. high. $1,800 £900

A pair of Chinese cloisonne enamel three-branch candelabra, 42cm. high. $2,400
£1,200

19th century French bronze torchere, in the form of a boy holding an elaborate eight-light mount.
          $7,800 £3,900

A pair of Louis XVI ormolu mounted candelabra, attributed to Pierre Gouthiere, 2ft.7½in. high.
          $8,000
£4,000

A pair of bronze and ormolu figures of a Bacchante and companion, in the manner of Clodion.
          $8,800 £4,400

A pair of 19th century candelabra made by Emile Guillemin, Paris, 9ft. 9in. high. $58,000 £29,000

CANDLESTICKS

19th century ornate, bronzed soft metal, candlesticks on circular pierced bases. $50 £25

Pair of Victorian spelter figures, 12ins. high. $90 £45

Pair of embossed bronze candlesticks entwined with dragons.
$100 £50

Pair of early 19th century French ormolu candlesticks with detachable sconces, circa 1810, 11¾ins. high.
$130 £65

Pair of 19th century ormolu, rococo style French candlesticks, 12ins. high. $190 £95

A pair of French ormolu candlesticks with figures of children, on shaped bases, 8ins. high. $220 £110

Pair of 19th century bronze and
cloisonne candlesticks. $320 £160

A pair of early 19th century
Angel candlesticks. $340 £170

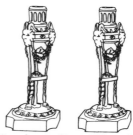

Pair of Regency ormolu and
bronze candlesticks. $360 £180

One of a pair of Louis XVI ormolu
candlesticks on white marble
plinths, 9in. high. $1,300 £650

Pair of bronze pricket candlesticks,
15th century. $2,000 £1,000

Pair of Tiffany bronze and glass
candlesticks, 14½in. high.
                    $3,000 £1,500

Small 19th century bronze model signal cannon. **$120 £60**

A well made bronze starting cannon with a turned barrel, 17in. long. **$150 £75**

Heavy bronze mortar of squat form, weight approximately 20lbs., base diameter 2ins. **$180 £90**

A good old model field cannon in t[ ] style of the late 17th century, 11in. barrel. **$200 £1**

A fine old model of a cannonade with heavy stepped bronze barrel, 8½in. long. **$220 £110**

An 18th century Malayan deck can[ ] Lantaka, 26½in. long, of cast brass. **$220 £1**

42

A cast bronze Malayan swivel cannon Lantaka, of good brown patina, with 32in. trunnions, mounted on a wooden stand. $310 £155

A good 18th century, bronze field cannon, the breech mounted with a coat of arms, on original wooden carriage, overall length 52in. $430 £215

A fine and well made model of a 17th century German field gun with a cast bronze barrel, 16½in. long. $480 £240

Pair of C. Lucknow, Arsenal, small bronze howitzers, 12in. long. $780 £390

One of a pair of very attractive bronze Georgian signalling cannon, with stepped oak carriage and wooden wheels, circa 1828. $2,200 £1,100

One of a pair of French bronze cannon inscribed 'La Prude' and 'La Juste'. $3,000 £1,500

CASKETS

Japanese bronze oblong box with
bird head handles, 9½ins. wide.
$80 £40

A shaped casket with ringed lion
carrying handles, and standing on
four claw and ball feet, circa 1750,
8ins. wide, 3½ins. deep, 4½ins.
high. $150 £75

An Italian black metal casket
with hinged cover, 13¼ins.
wide, 12½ins. high. $220 £110

Mid 19th century French Second
Empire ormolu jewel casket,
signed C. De Franor, 10ins. wide.
$340 £170

17th century money box.
$1,400 £700

Russian silver gilt and cloisonne
enamel casket, 5ins. wide.
$3,000 £1,500

44

20th century oxidised metal electric chandelier of three lights. $120 £60

A bronze chandelier with six scroll branches, fitted for electric light.
$200 £100

Attractive five branch brass chandelier, circa 1850.
$280 £140

19th century French rococo style brass chandelier, 21ins. wide.
$600 £300

A crisply cast nine-light ormolu chandelier, Paris 1850. $1,500 £750

A superb Adam style chandelier, circa 1785. $13,000 £6,500

CHERUBS

A pair of 19th century bronze cherubs on wooden bases, 20.5cm. high.
$150 £75

Pair of 19th century French bronzes depicting cupids at play. $200 £100

Bronzed metal figure of Cupid, signed Palcomet, circa 1870, 1ft. 4½in. high. $280 £140

Pair of mid 19th century French bronze cupids, 8in. high. $300 £150

Bronze figure of Cupid, signed Auguste Moreau. $640 £320

A pair of 19th century French bronze cupids with basket of flowers, 14in. high. $940 £47

Bronze entitled 'Enfants au Cover' of two entwined cupids wrestling for a heart, by Falconnet. $1,100 £550

Pair of French bronze putti on chased ormolu bases, 27cm. high. $1,200 £600

Pair of early 19th century French bronzes with marble bases, 16in. high. $1,200 £600

One of a pair of mid 18th century French bronze figures of putti. $2,800 £1,400

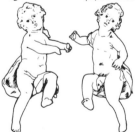

Pair of Louis XV gilt bronze putti by Philippe Caffieri, 13in. high. $3,200 £1,600

A very fine figure of the sleeping infant St. John, believed to be of 17th century Flemish origin. $13,000 £6,500

Seated bronze Buddha, on 3-tier pedestal, black lacquer overall, with red and gold decoration. $90 £45

Chinese bronze group of a figure and a cow, 8½in. wide. $100 £50

19th century Chinese bronze of an old man seated on a large fish supported by stylised waves, 8½in. high. $110 £55

18th century Chinese bronze Buddha on a wooden stand, 27cm. high. $130 £65

Chinese bronze statuette of a deity in ceremonial dress, 14in. high. $170 £85

A heavy 18th century Chinese bronze Buddha, 10in. high, with traces of gilt. $290 £145

One of a pair of early Chinese dogs of Fo. $1,150 £575

18th century bronze and Chinese enamel figure of Kwan Yin, 56cm. high. $1,250 £625

Early 15th century Ming bronze figure of Golden Boy, 6.5in. high. $1,360 £680

Chinese openwork bronze disc of the 6th-5th century B.C., 2¾ins. diam. $3,800 £1,900

Shang dynasty archaic bronze wine vessel. $3,800 £1,900

Bronze statue of a cat from the 26th Dynasty. $6,600 £3,300

Chinese gilt bronze of the Hau Dynasty, depicting a bear seated on its haunches, 9.8cm. high.
$32,000 £16,000

14th century Chinese gilt bronze figure of Guatama, 20cm. high.
$16,000 £8,000

Gilt bronze sea dragon from the period of the six dynasties, 5½in.
$150,000 £75,000

Chinese, Shang/early Western Chou dynasty, bronze pole finial, 15cm. high.
$160,000 £80,000

T'ang gilt bronze figure of a striding lion, 6½in. high.
$200,000 £100,000

A Shang dynasty bronze Fang-I.
$360,000 £180,000

Bronze figure of a discus thrower, 8½in. high.   **$60 £30**

A bronze reproduction, after the antique, of a satyr, the bearded figure running, arms raised, 1ft. high.   **$90 £45**

Bronze classical male figure, 13¼ins. high.   **$100 £50**

One of a pair of French gilt spelter female figures with cupids, 48cm. high.   **$100 £50**

Classical, deep green bronze figure of a satyr on stepped square base of bronze, 12½ins. high. **$120 £60**

A bronze statuette of Eve, standing with serpent entwined, on circular rockwork base, unsigned. **$150 £75**

Bronze female winged statuette, stamped J. S. Westmacott, dated 1852. $150 £75

Late 19th century bronze figure of Ceres, after Fulconis. $150 £75

A pair of 19th century bronze figures of Diana and a male god, on square modelled plinths, 8in. high. $150 £75

A bronze statuette of Mercury, seated on a rockwork base, 12in. high.
        $150 £75

A bronze mythological female winged statuette on ball and square alabaster plinth base, 27in. high. $200 £100

19th century bronze figure of a maiden on a white marble base, overall height 3ft. 4in. $200 £100

52

Pair of bronze figures, circa 1850.
$260 £130

19th century French bronze figure of
'The Winged Mercury' signed Clodion,
2ft. 9in. high. $350 £175

Bronze study of a seated muse on
ebonised plinth, 12½in high.
$480 £240

A pair of bronze statuettes of Mercury
and Minerva on marble plinths with
bronze band of figures. $460 £230

19th century bronze figures of
Mercury and Juno. $530 £265

Bronze group of a centaur attacking
a warrior, 21in. high. $380 £290

53

Well modelled bronze of Venus, 34in. high, circa 1860. $790 £395

Bearded nude man carrying a dead boar, bronze, 16½in. high, on rouge marble plinth. $840 £420

Bronze figure of Peace by Edward Onslow Ford, 22in. high. $900 £450

A fine bronze bacchante group with Pan and satyr, signed and dated Clodion, 1862, 23½ins. high. $900 £450

19th century Italian bronze figure of Cupid, 3ft. 1in. high. $900 £450

Late 19th century French bronze of Hercules, 14in. high. $960 £480

Bronze figure of Diana by Louis-Ernest Barries, late 19th century, 17in. high. $1,080 £540

17th century French bronze figure of 'Sleeping Ariadne' draped in classical robes, 22in. long. $1,800 £900

19th century bronze figure of Ulysses, 35in high.
      $1,800 £900

One of a pair of ormolu chenets of Louis XV style, 1ft.3in. high.
      $1,900 £950

Late 16th century Italian bronze statuette of Bacchus, on a giltwood base, 8¾in. tall. $1,920 £960

Bronze figure of Juno by Antoine-Louis Barye, 11in. high.
      $3,000 £1,500

Late 18th century pair of Italian bronze centaurs, attributed to Giacomo Zoffoli on griotte marble bases, 11in. tall. $4,000 £2,000

A Greek bronze griffin of 7th-6th century B.C., 5¾in. high. $5,000 £2,500

A 19th century life-size bronze figure of 'David' after the original by Donatello. $6,000 £3,000

Pair of Padvan bronze candlesticks from the workshop of Andrea Riccio. $6,600 £3,300

'Apollo and Daphne' after the marble by Giovanni Lorenzo Bernini, 1591-1680, 79cm. high. $8,800 £4,400

'Pluto and Proserine', a large bronze after Bernini, 30½ins. high.
$8,800 £4,400

Bronze half size replica of 'The Little Mermaid' by Edrard Erikson.
$9,500 £4,750

17th century Italian bronze of Nessus and Deianira, 44.5cm high.
$36,000 £18,000

Germano-Italian copper-gilt figure of David with his right foot resting on the head of Goliath, 19.8cm. high, circa 1600. $80,000 £40,000

Pair of bronze statues of Venus and Mars by Tiziano Aspetti, 21in and 21½in high. $130,000 £65,000

57

Victorian dog doorstop. $40 £20

A cast bronzed dog door stopper.
$50 £25

19th century bronze of a dog
curiously surveying a tortoise, by
A Jacquemart. $200 £100

19th century unsigned bronze of two
dogs nuzzling the tall grass for their
quarry. $240 £120

Victorian bronze figure of a dog.
$260 £130

A bronze model of a retriever with
its right paw raised, signed P. J.
Mene, mid 19th century, 4½in.
long. $300 £150

19th century bronze figure of
a greyhound, 6ins. tall.
$380 £190

19th century bronze of a greyhound
and bird. $380 £190

French bronze bulldog on a square
base, 11in. long, signed P. Dinby,
circa 1900. $400 £200

A bronze whippet, signed P.J. Mene.
$530 £265

A bronze whippet by P.J. Mene.
$530 £265

A French bronze figure of a stretching
dog, by E. Fremiet, 3¾ins. high.
$550 £275

19th century bronze of gundogs signed L. Bureau, 14ins. high, 18ins. long. $600 £300

A bronze retriever with pheasant in mouth, 15½in. high, signed J. Moigniez. $720 £360

An unusual bronze group 'Setter, Pointer and Pheasant' stamped Barye, 10½in. long, 7½in. high. $770 £385

Bronze of a foxhound 'Forager' by Adrian Jones, 10ins. long, 10½ins. high. $840 £420

19th century bronze of two whippets by P.J. Mene. $840 £420

Bronze figure of a whippet and a King Charles spaniel, 6ins. high, 10ins. long by P.J. Mene. $860 £430

A French bronze figure of a pointer with head raised, by P.J. Mene, 8ins. high. $890 £445

A French bronze group of two pointers with a dead hare, signed P. J. Mene, dated 1872, 8in. long. $1,040 £520

Bronze study of two dogs approaching a grouse, signed P.J. Mene, 1847. $1,060 £530

One of a pair of whippets by P.J. Mene. $1,100 £550

19th century bronze figure of a retriever signed E. Wunshe, 1ft.9in. high. $1,800 £900

One of a pair of 19th century French lifesize bronze Great Danes by Georges Gardet, 3ft.4in. high x 4ft.4in. long x 1ft.7in. wide. $10,000 £5,000

A bronzed spelter group of a warrior on horseback, on an ebonised plinth base, 18ins. high. $60 £35

One of a pair of late Victorian spelter Marli horse figures $90 £45

A bronze group of a horse and negro boy, on ebonised base, 11ins. high. $100 £50

A bronze statuary group, of a man in armour on a charger, signed on base, 'A. Dodds, 1949, 15¾ins. high. $180 £90

19th century bronze model, 26in. high. $210 £105

Very handsome, hollow backed, gilded bronze classical figure of a woman on horseback, signed 'Coutan', 20ins high, 17½ins. long. $290 £14

A farmer on horseback, harrowing signed 'E. Drouot', on a rouge marble base, 17ins. wide. $410 £205

One of a pair of English bronzes depicting horses and jockeys, 5in. high, circa 1840. $410 £205

One of a pair of bronze groups of prancing horses with figure attendants, on oblong shaped bases, 19½ins. high. $410 £205

19th century bronze of a horseman by Celleoni Verrpochio. $410 £205

One of a pair of bronze groups of horses and attendants, on rockwork bases, signed Couston, 20ins. high. $460 £230

19th century bronze of a hunter on horseback, signed E. Drouot, on a rouge marble base, 16ins. wide. $550 £275

63

A bronze of a jockey and horse, by
C. Sotyens, 1891. $600 £300

19th century bronze by Boehm.
$770 £385

Pair of large bronze Marli horse groups, signed C.H. Crozatier, 23ins. high.
$770 £385

Bronze of an Indian warrior entitled
'Appeal to the Great Spirit', signed
C.E. Dallin, 1913, 20ins. high.
$960 £480

A Western United States bronze
depicting the shoeing of an
immigrant farmer's horse, by
Carl Kauba. $1,010 £505

One of a pair of French 19th century bronze Marley horses, 24in. high. $1,080 £540

19th century bronze horse and rider, signed on base, 26ins. high, 22ins. wide. $1,340 £670

air of horsemen by T. and T.H. Gechter, 1841 and 1842, 14ins. high.
$1,340 £670

A bronze group by J. Willis Good, signed and dated 1875.
$1,380 £690

A large French bronze group with female figure mounted on a horse, signed on base, L. Chalow, 32ins. high. $1,440 £720

65

EQUESTRIAN FIGURES

Bronze equestrian group by Godefroid Devreese, 23ins. high. $1,490 £745

Bronze equestrian group by Jules Moigniez. $1,500 £750

Bronze figure of 'Jockey and Racehorse' by Moigniez.
$1,560 £780

19th century knight on horseback in bronze, ormolu, ivory and onyx
$1,560 £780

19th century bronze of a horse and jockey signed P. J. Mene, 17in. high.
$1,600 £800

One of a pair of fine brass and bronze Marli horses on ormolu mounted boulle plinths, 28ins. high. $1,620 £810

Bronze group of Russian soldiers, 14½in. high. $1,700 £850

A bronze group of three Dutch boys climbing on to a cart horse, inscribed by E. Martin, Fondeur, Paris.
$2,040 £1,020

19th century bronze cavalry charger and trooper by J. Willis Good.
$3,000 £1,050

One of a pair of equestrian bronzes by J. Willis Good. $2,100 £1,050

A bronze racehorse and jockey, signed I. Bonheur, circa 1860, 18½in. long. $2,100 £1,050

Early 20th century bronze racing group, 12 x 38in., signed Yves Benois Gironier. $2,400 £1,200

67

Mid 18th century pair of large equestrian figures of Turkish warriors.
$2,400 £1,20

Bronze 'Trotting Horse' by Count Grimaldi, 1886, 29ins. long.
$2,400 £1,200

Russian bronze group by Lancera
$2,500 £1,250

Bronze group of a jockey astride a racehorse by Pierre Jules Mene.
$3,520 £1,760

A large bronze figure of a 'Raceho
and Jockey' signed I. Bonheur.
$3,520 £1,7

A bronze seated female figure, 8in. high. $50 £25

A pair of French bronze spelter female figures of 'Gaiete and Modeste', on ebonised plinths, 12in. high. $80 £40

Victorian bronze-coloured spelter figure, 20in. high. $90 £45

A bronze statuette of a lady wearing a long dress and holding a fan, on an alabaster base, (unsigned), 15in. high. $130 £65

A bronze statuette of a lady with her right arm outstretched, on a circular base, unsigned. $160 £80

Victorian bronze figure of a female on a marble base, 21in. high. $170 £85

FEMALE FIGURES

Bronze statuette of Joan of Arc in full armour, 9in. high.
$180 £90

One of a pair of Victorian bronze female figures, 14in. high.
$180 £90

A bronze study of a slender young woman, 10in. high. $190 £95

19th century bronze figure depicting a windswept girl, unsigned, 19in. high
$200 £10

Bronze figure of Daphne by Josephine Sykes, 1930, 22½in. high. $240 £120

Bronze figure of a woman by Charles Sykes inscribed 'Phryne-Women', 19
$240 £1

An interesting bronze of the 'Iron Maiden', with hinged and sprung doors which open to reveal a gilt and lacquered figure of a nude woman, 10in. high. $290 £145

Standing bronze figure by O. Vabacchi Topino. $380 £190

9th century bronze nude, 26in. high. $380 £190

Semi nude bronze female figure by E. Drouot, 28in. high. $400 £200

French gilt bronze figure of a young girl, after Gustave Obiols. $400 £200

A German bronze statuette of Venus de Milano, on an oblong base, stamped J. Waltz, 3ft. 7in. high. $470 £235

FEMALE FIGURES

One of a pair of spelter figures after Moreau, circa 1900, 24½in. tall. $500 £250

French, 19th century, bronze fountain figure on a Carrara marble base, overall height 38in. $540 £270

Italian bronze figure of a dancer signed G. Beneduce on a marble plinth, 73cm. high. $580 £290

Art Deco ivory and bronze figure by Gregoire. $600 £300

A bronze statuette by Gregoire entitled 'La Charite'. $640 £320

Late 19th century French bronze, depicting a traditionally dressed Nubian woman playing the harp. $640 £320

Bronze study of a native girl,
signed Salmson, 53cm. high.
$720 £360

Elaborate bronze figure of an
Egyptian priestess. $900 £450

Fine quality bronze by L. Madrassi,
29ins. high. $960 £480

Signed Carrier gilt bronze figure of
a mandolin player. $1,580 £790

'La Baigneuse Frileuse' by Jean
Baptiste Carpeaux. $3,080 £1,540

Bronze 'Mother and child in a rocking
chair' by Henry Moore, 13ins. high.
$53,130 £26,565

FIGURE GROUPS

An African bronze group of a man and woman seated, 5in. high.
$60 £30

A Malayan bronze group of a man and woman , 5in. high. $60 £30

Pair of Victorian bronzed spelter figures on marble bases, 14in. high.
$100 £50

A large pair of spelter figures of a blacksmith at his anvil and a farmer with ploughshare. $160 £80

19th century French spelter group 'La Fete des Fleurs' by Aug.Moreau.
$180 £90

A pair of French gilt bronze figures, 'Une Repitition, par Faure de Brousse', on green onyx and gilt metal bases, 12in. high. $190 £95

A pair of bronze figures of boys playing cricket, signed and dated 1863.
$240 £120

A primitive Yoruba bronze group of crude construction, and uncertain date, containing eight separately made figures, base, 6in. x 4½in. $240 £120

19th century bronze of the Laocoon.
$240 £120

19th century pair of bronze figures of Africans, 11½in. and 12½in. high.
$310 £155

One of a pair of bronze figure bookends on slate bases. $320 £160

Pair of 19th century African bronze figures. $360 £180

75

**Bronze of two duelling musketeers
by E. Drouot, 17½in. long.**
$720 £360

**Pair of French bronze figures of
Cavaliers, 21in. high.**
$840 £420

**A 19th century French bronze group
of lovers, 61cm. high.** $960 £480

**18th century bronze figure group
signed Clodion, 16½in. high.** $960
£480

**Italian bronze group by A.
Pandiani of Milan, circa 1880.**
$1,040 £520

**Pair of 18th century bronze statuettes
of gentlemen on yellow marble plinths,
20in. high.** $1,080 £540

Egyptian bronze figure of Isis from the late Dynastic period. $1,080 £540

19th century bronze figures, 20in high. $1,400 £700

Canadian bronze group 'Le Premiere Baiser' by Alfred Laliberte. 46cm. high. $1,860 £930

Late Victorian bronze group by E. Piat, 20in. long. $1,950 £975

Mid 19th century rare bronze corkscrew, possibly French. $2,100 £1,050

17th century gilt bronze of 'The Three Graces' by George Petral. $32,000 £16,000

FIGURES WITH ANIMALS

A bronze soldier and dog, circa 1840.
$130 £65

A classical bronze group of a goat and
two figures of children, 26cm. high.
$500 £250

A large bronze mythological group of
Europa the bull and four other figures,
on square shaped base, with deer and
boars, 16in. high. $720 £360

Bronze group of a highlander with a
pair of wolfhounds by P. J. Mene,
circa 1880, 23½in high.
$1,240 £620

Bronze of an African horseman being
attacked by a lion, by Isadore Bonheur.
$1,440 £720

Bronze statue, 3in. tall, of an infant
with cockerel, signed G. Marty, base
inscription Adriano Cecioni, Firenze.
$2,200 £1,100

19th century horse bronze of good colour. $300 £150

Bronze study of a stallion, signed P.J. Mene, 126cm. high.
                                              $550 £275

19th century bronze model of a stallion, 12in. high. $1,040 £520

Bronze figure of a hunter by J. Willis Good. $1,200 £600

Bronze of an Arab stallion 'Ibrahim' by P.J. Mene. $1,200 £600

Mid 19th century bronze figure of a racehorse, 19in. high.
                                              $1,360 £680

Bronze of a horse on a marble base, 22cm. long. $1,800 £900

Bronze figure of an 'Arab Stallion' by Moigniez, 15in. long, 12½in. high. $1,800 £900

An equestrian group by Christopher Fratin. $1,800 £900

One of a pair of bronze horses by Mene. $1,920 £960

A bronze group of a mare and stallion by J. Moigniez, 16in. high. $1,920 £960

A bronze mare and foal by Isadore Bonheur. $1,920 £960

80

Fine bronze of a mare and stallion by Monier.   **$2,400 £1,200**

A fine bronze 'Mare and Foal' by Fratin.   **$2,640 £1,320**

A bronze entitled 'L'accolade' by P.J. Mene.   **$3,410 £1,705**

Superb bronze of a Turkish horse by Barye, 11¾in. high. **$8,800 £4,400**

A late 16th century/early 17th century bronze figure of a cheval ecorche, attributed to the Florentine Bolgna-Susini Workshops, 92cm. high.
    $310,000 £155,000

Bronze figure of a horse by Giovanni Bologna.   **$12,000 £6,000**

81

INCENSE BURNERS

19th century bronze vase with pierced cover, 6ins. $30 £15

A Chinese 19th century bronze incense burner, the domed lid cast and pierced with entwined dragons, 12ins. high.
$120 £60

18th century Chinese bronze incense burner. $140 £70

19th century Kylin bronze incense burner, 7½ins. high. $190 £95

19th century lacquered bronze incense burner. $300 £150

Early 19th century cloisonne enamel incense burner. $500 £250

One of a pair of 19th century clois-onne incense burners in the form of dragons, 7in. long. $650 £325

One of a pair of 19th century Chinese cloisonne, elephant shaped, incense burners, 5½in. high. $720 £360

Mid 19th century fish incense burner, fashioned in enamelled silver with gilt mounts, 20in. tall. $1,150 £575

Pair of 18th century cloisonne enamel incense burners modelled on standing quail, 5ins. high. $1,450 £725

Pair of Ch'ien Lung cloisonne incense burners, each supported on gilt bronze elephant heads, 26in. high.$1,800 £900

Pair of Chinese Ch'ien Lung cloisonne enamel incense burners in the form of quails, 5¾ins. high. $2,400 £1,200

INDIAN BRONZES

A 19th century Indian figure of Krishna and fool in bronze.
$150 £75

19th century Indian bronzed brass female deity, 17ins. high. $180 £90

Indian bronze group of chieftain and three attendants on an oval cobra pattern base. $240 £120

An old heavy Southern Indian bronze standing figure of the Celestial Maiden probably circa 1500 or earlier, 16½ins. high. $420 £210

A rare Samaskanda group, Indian, 11th-13th century. $7,700 £3,850

11th century Chola bronze figure of the god Siva-Nataraja.
$96,000 £48,000

Art Nouveau solid bronze inkwell, circa 1900, 7in. long. $48 £24

A bronze cupid inkwell, circa 1830.
$110 £55

French bronze inkwell, 5in. high, 7½in. wide. $110 £55

Bronze model of an owl perched on inkwell stump. $150 £75

Automobiles Ferman', a bronze figure f Daedalus on a marble base fitted ith two inkwells by G. Colin, 1907.
$770 £385

16th century Venetian bronze inkwell. £1,400 £700

A bronze group of a cat and frog on circular base, unsigned, 8ins. high, 6ins. diam. $100 £50

An ingeniously made, mid 19th century Japanese, articulated bronze grasshopper. $100 £50

A good 19th century Japanese tiger bronze, roaring with head upturned. $200 £100

Japanese bronze standing figure of a warrior in a fighting pose, 18ins. high. $240 £120

19th century Japanese bronze figure of a warrior on a carved hardwood base. $240 £120

A fine Japanese figure of a warrior wearing his Daisho of swords, he draws his Itatina, 10½ins. high. $300 £150

Japanese bronze figure of a man.
$1,080 £540

An early Japanese bronze figure,
6ins. high. $1,100 £550

A large 19th century Japanese bronze
group. $1,200 £600

19th century Japanese bronze
figure of a Samurai warrior, 10in.
high. $1,320 £660

Japanese red bronze box, signed
Yoshiaki, 10.8cm. $2,000 £1,000

Large Japanese bronze model of a
tiger, 6ft.6in. long. £4,400 £2,200

JARDINIERES

19th century Japanese bronze circular jardiniere with bird and dragon in relief, 26½cm. high. $90 £45

Chinese brass and cloisonne enamel circular jardiniere, 12in. diam.
$120 £60

Chinese oval cloisonne jardiniere, 10in. wide, 7in. high. $170 £85

Japanese bronze circular jardiniere with birds and flowers in bas relief, 40cm. diam. $220 £110

Japanese cloisonne enamel jardiniere, 32cm. diameter.
$280 £140

A square onyx jardiniere with cloisonne enamel mounts, 10in. wide. $300 £150

19th century electric lamp, the base with a bronze eagle. $40 £20

Victorian bronzed spelter figure lamp, 18½ins. high. $80 £40

Pond lily table lamp in the Art Nouveau style. $100 £50

Bronze 19th century candelabrum with green glass shades, 15in. high. $120 £60

A bronze-framed model of a church, the sides inset with lithophane panels of church and country scenes.
                   $130 £65

19th century magician's lamp cast in bronze, 7½in. tall. $150 £75

An elegant bronze Art Nouveau lamp, 15in. high. $160 £80

A very handsome Victorian oil lamp made in bronze and Paris porcelain, circa 1860. $190 £95

Taperstick by Christian Hammer in the form of a sandalled foot, Stockholm 1857, 5¼in. long.
$200 £100

A pair of French, electric table lamps, ormolu, with bronze stems in the form of children seated on gilt metal tortoises, 13ins. high. $280 £140

A large 'Wooden Wall' ship style lantern, 45in. high. $280 £140

George III bronze-framed porch lantern. $960 £480

Tiffany lamp with a gilt bronze base and coloured glass poppy pattern shade. $2,000 £1,000

Bronze lamp in the style of Gurschner. $1,800 £900

Louis XVI ormolu wall light, 1ft. 9in. high, one of a pair.
$3,000 £1,500

A Jugendstile bronze lamp of Loie Fuller by Raoul Larch. $3,000 £1,500

A bronze and shell lamp by Gurschner.
$4,000 £2,000

Tiffany Studio wisteria lamp. $32,000 £16,000

MALE FIGURES

Small late Victorian spelter figure of a cricketer. $30 £15

Oriental bronze figure of a man, on a wooden base, circa 1890, 9in. high. $70 £35

A Victorian gilt bronze figure of a man, on marble base. $70 £35

West African cast bronze figure of a seated smoker, 6½in. tall. $80 £40

A French bronze statuette 'Arabe en Marche', on octagonal base, 13in. high. $80 £40

Bronze statuette of King Henry VIII on black marble square base. $90 £45

19th century bronze of a youth removing a thorn from his foot, 8ins. high. $120 £60

Bronze figure of Harry Vardon by Henry Pegram, circa 1908, 14in high. $170 £85

19th century bronzed brass figure of a fisherman, on a marble base, 12¼in. tall. $180 £90

Bronzed brass figure of a farmer, on a shaped black marble base, 12in. tall. $180 £90

A Continental bronze figure of a man seated on a rockwork base, signed A. Filepovie, 12½in. high. $200 £100

Finely modelled French bronze figure, 14in high, circa 1870. $230 £115

16in. tall bronze figure of an athlete, signed Schmidt Felling, 1905.
$280 £140

Victorian bronze of a drunkard.
$470 £235

19th century Italian bronze figure of an athlete, by Canova. $500 £250

Bronze male statuette 'Le Boucheron', signed Chambard, 34in. high.
$480 £240

19th century French bronze of an Artist. $600 £300

Large 19th century bronze figure, 28in. high, unsigned. $600 £300

Late 5th century B.C. Etruscan bronze figure of a warrior. $600 £300

Bronze Florentine lute player by A. E. Gaudez, 18in. high.
    $700 £350

Late 19th century French bronzed spelter figure of a Harlequin, 38in. tall. $700 £350

A large bronze figure of a man tearing apart a tree trunk, after the antique, on square base, 31in. high. $720 £360

Late 19th century bronze figure of a Moroccan falconer, 24in. high.
    $840 £420

'The Sluggard', bronze circa 1890, after Lord Leighton's sculpture at the Royal Academy, 1886, 1ft.9in. tall. $960 £480

One of a pair of 19th century male bronzes, signed Salmson.
$1,200 £600

Bronze figure of a shepherd, from the circle of the Italian master Giovanni Bologna, 4½in. tall.
$1,320 £660

Bronze Indian fisherman, 26¾in.
$1,440 £720

Bronze of George III by Lawrence Gahagan, 11in. high. $1,600 £800

Bronze figure of 'The Sluggard' by Lord Leighton, 1886, 17in. high.
$1,900 £950

Bronze figure of a miner by Constantin Meunier, 22½in. high.
$2,500 £1,250

Bronze of 'Figaro' by Jean Baptiste Carpeaux. $3,300 £1,650

'The Mieliepap Eater', a bronze by Anton Van Wouw (1862-1945), 14cm. high. $4,400 £2,200

ronze entitled 'La Berie de la Danse', y Jean Baptiste Carpeaux.
$5,000 £2,500

One of a pair of bronze athletes, 4ft. 9in. high. $5,400 £2,700

Late 16th century German bronze, 33cm. high. $11,000 £5,500

Gilt bronze figure of St. John the Evangelist, 3¾in. high, circa 1180.
$76,000 £38,000

97

An old African brass amulet in the form of a mask, 4¼in. high.
$60 £30

Bronze belt mask from Benin, Niger. 6in.
$100 £5

19th century bronze devil mask of Lucifer with a rasping tongue, 9in. high.
$150 £75

Ogoni mask from south-east Nigeria 24cm. high.
$840 £420

Nigerian bronze panther's head mask.
$2,000 £1,000

'La Muse Endormie', a bronze by Constantin Brancusi, 10½in. high
$150,000 £75,000

A bronzed spelter figure of a warrior, in armour, 17in. high. $70 £35

French bronze statuette 'Janissaire' Prix de Rome, on oblong base.
$120 £60

A French bronzed metal statuette of 'Victorieux   Jour de Gloire', on a green marble base, 33in. high.
$120 £60

Bronzed brass figure of Napoleon wearing a military tailcoat with Orders, 7in. high. $130 £65

pair of French bronzed spelter ures of 'Le Prince Noir' and 'Le Roi n', in suits of armour, on circular nised plinths, 25in. high. $130 £65

19th century bronzed and gilded spelter figure of the Duke of Wellington. $130 £65

MILITARY FIGURES

Bronze figure of an 18th century French standard bearer, 9½in. high, wearing tailcoat, cocked hat and breeches. $150 £75

Bronze and gilded spelter figure of Lord Nelson, 11in. high. $150 £75

A good bronze standing figure of an Imperial German Jaeger soldier, 9¾in high on heavy marbled base.
                                          £160 £80

Boer War bronze of 'A Gentleman in Khaki', 9in. high. $170 £85

Miniature suit of armour in the late 15th century Maximillian style, 17in. high. $180 £90

A 19th century cast bronze figure of 'Polish Infantryman', 10½in. high.
                                          $180 £90

niature suit of armour in 16th cen-
ry style, with articulated arms,
½in. high. $220 £110

One of a pair of bronzed brass
figures of Charles I and Oliver
Cromwell, 13in. high. $240 £120

good standing bronze of a Franco-
ussian period German Jaeger Rifle-
n, 12½in. high. $240 £120

Imperial German presentation bronze
statue of a standard bearer in the
Kaiser Franz Garde Grenadier Regi-
ment. $260 £130

A standing bronze of Lieut. von
chmelling at the Battle of Ligny,
wearing full uniform with drawn
abre. $280 £140

19th century bronze of a soldier in
18th century uniform, 12in. high.
                             $300 £150

101

MILITARY FIGURES

A well made model suit of jousting armour in the style of 1530, 15½in. high. $300 £150

A well executed bronze figure of a German mercenary in the style of circa 1600, the base signed Deniere and stamped H. Picard. $420 £210

Bronze figure of a Grenadier Guard, 1889, on plinth, signed G.E. Wade, stamped H. Luppens and Co., 18in. high. $670 £335

Japanese signed bronze of a warrior carrying a Yari. $300 £1

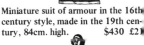

Miniature suit of armour in the 16th century style, made in the 19th century, 84cm. high. $430 £2▶

19th century French bronze of a swordsman. $670 £335

First World War figure in bronze
by R. Tait McKenzie, 1916, 16½in
high. $900 £450

Pair of French bronze figures of warriors in armour, on circular bases,
signed A. Garrier, 24in. high.
                          $960 £480

Pair of 19th century bronze studies of
Eastern warriors, signed Deniere, 14½in.
high. $1,010 £505

Fine bronze of a Red Indian by
Carl Kauba, on marble plinth, 19th
century, 16in high. $6,800 £3,400

Late 18th century pair of coloured
bronze models of Red Indian chiefs by
C. Kauba. $9,000 £4,500

Bronze by Henry Moore entitled
'Warrior with Shield'. $13,000 £6,500

103

A 19th century Chinese bronze rect-
angular vessel, with tapering sides cast
with panels of Chinese fret and dragon
masks, 7½in. long. $60 £30

A good quality bronze skillet.
                                    $84 £42

Chinese bronze wine vessel on triple
legs. $100 £50

Standard bronze measure with con-
tainer, 1601. $480 £240

Set of ten gun-metal standard capa-
city measures from bushel to quarter
gill. $2,400 £1,200

A pair of fine cloisonne enamel circular trays of the Ch'ien Lung period.
                          $1,680 £84

Oriental bronze ewer with figures and animals in relief. $50 £25

A good late 18th century cloisonne enamel teapot. $1,500 £750

Persian metal figure of a simurgh, 19th century, 48cm. high. $2,200 £1,100

An original Nazi standard-top eagle, with swastika within wreath, 10in. high. $150 £75

Mid 19th century gilt bronze fire guard, 66cm. high. $320 £160

A pair of 19th century classical bronze ewers. $1,200 £600

105

One of a pair of bronze cheek-pieces from a horse's bit, Iranian, 14cm. square. $17,600 £8,800

One of a pair of Thai bronze drums, 48cm. high. $1,600 £800

Pair of large mid 19th century bronze firedogs. $740 £370

One of a pair of Louis XVI ormolu chenets on leaf-cast plinths, 1ft. high. $1,000 £500

19th century ormolu birdcage with ogee top, 24in. high. $2,800 £1,400

1st century B.C. Provincial Greek bronze foot. $720 £360

A Provincial Greek bronze hand of the 1st century B.C. $480 £240

Early 19th century French alarm mechanism, in bronze case, standing on lion's paw feet, 5.8cm. high.
$780 £390

Russian tea service in cloisonne, made in the 1890's.
$17,000 £8,500

A Japanese, bronze, oval, double-handled hibatchi with pierced wood cover and stand, 10ins. high. $30 £15

An unusual book carrier of finely figured oak veneer on a mahogany base with a Victorian Gothic bronze gallery. $180 £90

A fine Queen Anne period apothecary's bronze mortar with unusual decoration of thirteen raised bosses around band, circa 1710, 4in. diam., 3in. high. $90 £45

Bronze Fleur de Lys pestle and mortar. $110 £55

English bronze mortar and pestle, circa 1690. $220 £110

Continental pestle and mortar embossed MDL. XXXIV. $260 £130

17th century bronze mortar with the inscription 'William Boult, 1654'.
                                   $860 £430

Large bronze mortar by Luke Ashton of Wigan. $900 £450

An old Nepalese seated figure of Buddha, with hollow base, 8in. high.
$110 £55

18th century Nepalese bronze of Hayagrava in Yab-Yum, 6¼in. high.
$900 £450

17th century Nepal bronze of the Bodhisattva Avalokitesvara, 5½in. high.
$1,440 £720

18th century Nepalese copper and silver figure of Vajravarshi, 6¼in. high.
$1,800 £900

Nepalese gilt bronze figure of Maitreya, 25½in. high.
$8,000 £4,000

A 14th century Nepalese gilt copper figure of Guatama seated in dhyanasana, 58.5cm. high. $54,000 £27,000

PLAQUES

A bronze classical female head door mount with bas-relief vines. $16 £8

Art Nouveau, bronze wall panel, 8½in. x 5¾in. $50 £25

19th century bronzed copper plaque in a carved oak frame, 16¾in. x 7½in. $100 £50

A bronze plaque of Napoleon, circa 1850, 10in. diam. $100 £50

19th century Japanese cloisonne dish, 14in. diameter. $110 £55

Large 19th century Japanese cloisonne enamel dish, 36in. diam. $1,490 £745

A fine bronze sundial with circular base plate engraved with the points of the compass, hours and minutes, signed by Nairne & Blunk, London, 1809, 10in. diam. $80 £40

Heavy sheet bronze vertical wall sundial, the bronze Roman numerals soldered onto the plate, 12in. square. $160 £80

Octagonal incised slate sundial, 10½in. across, with cast bronze gnomon. $160 £80

An exceptionally fine horizontal sundial in engraved bronze, the plain gnomon with original strengthening bars either side, inscribed 'Cole Fecit', 10in. diam. $230 £115

17th century brass sundial by Henry Wynne, London, 16¾in. diameter. $700 £350

Fine universal sundial by Watkins and Hill. $700 £350

THAI

19th century bronze Thai Buddha seated on a stepped throne. $130 £65

A bronze deity seated playing a Koto, 11¾in. high, late 19th century. $340 £170

One of a pair of Thai bronze kneeling figures from the 19th century, 79cm. high. $1,000 £500

Bronze Thai Buddha, 30in. high. $1,100 £550

TIBETAN

Tibetan gilt bronze of a Lama seated on a carpeted podium, 6.5in. tall. $430 £215

West Tibetan bronze seated figure of Kubera holding a mongoose vomiting jewels, 3½in. high. $600 £300

16th century Tibetan bronze of Sherab Seng-Ge, 7¼in. high.
$840 £420

Tibetan bronze of an abbot, 6¼in. high, circa 1800. $1,080 £540

Tibetan bronze of Maitreya seated in Virasana with hands in Dharmacakra, 5in. high, circa 1600. $1,920 £960

18th century Tibetan bronze of Kubera, 8in. high. $1,920 £960

West Tibetan bronze figure of Kubera wearing jewellery, inlaid with silver and stones, 8½in. high. $3,500 £1,750

An 18th century Sino-Tibetan gilt bronze of the Hdarmapala Yamantaka in the form of Vajrabhairava, 57.1cm. high. $28,000 £14,000

113

Pair of Victorian bronzed soft metal urns, 8in. high. $50 £25

Small pair of marble and bronze late Victorian urns, 10in. high. $80 £40

A pair of European chinoiserie, ormolu and marble two-handled pedestal urns with covers, on paw feet, 10½in. high. $180 £90

Pair of early 19th century bronze urns, 9in. tall. $200 £100

A pair of fine ormolu mounted urns, the marble of a rare variety and colour with a tin stratum, 10½in. wide, 12½in. high. $720 £360

A pair of Regency bronze and white marble tazza, 10½in. high. $800 £400

114

Pair of urn-shaped bronze vases and
covers with bacchanalian scenes
in relief, 22in. high. $820 £410

A pair of neo-classical lapis lazuli and
ormolu cassolettes in the form of
vases, 8¾in. high. $1,110 £550

A pair of Louis XVI ormolu mounted onyx urns, 9in. high.
$1,300 £650

One of a pair of fine porphyry
tazze of oval form, 1ft. 5in. high.
$5,400 £2,700

Two of a set of three 18th century
Italian plant urns in gilded bronze.
$11,000 £5,500

A finely cast archaic bronze cauldron
of the Shang dynasty, 9¼in. high.
$34,000 £17,000

VASES

19th century Japanese bronze vase with Kylin handles, 6½in. high.
$30 £15

Pair of cloisonne enamel bottle-shaped vases with blue ground, 11in. high.
$76 £38

Pair of bronzed double handled vases with covers, on circular bases, 18ins. high.
$80 £40

A Oriental bronze double handled bottle shaped vase, inset with coloured enamel bands, 9½in. high.
$80 £4

Late 19th century enamel vase of yellow ground, 10in. high.
$100 £50

Pair of bronze double handled vases, 9in. high.
$120 £60

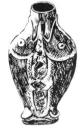

Pair of Japanese bronze vases decorated with flowering trees and birds, 9½in. high.　　　　$120　£60

A Japanese cloisonne enamel vase in the form of two fish, 14in. high.
　　　　　　　　　　　$120　£60

A 19th century Chinese bronze circular vase with repousse figures of birds, 10in. high.　$120　£60

Pair of cloisonne enamel vases with green ground, 7in. high.　$120　£60

Pair of 19th century bronze vases.
　　　　　　　　　$130　£65

19th century bronze and cloisonne baluster shaped two-handled vase, 18cm. high.　　　　$130　£65

117

VASES

19th century Japanese cloisonne enamel vase, 7in. high. $140 £70

One of a pair of Chinese bronze and coloured enamel hexagonal vases with double ring handles, 11in. $150 £75

One of a pair of bronze and enamelled barrel-shaped vases with ring handles, 12in. high. $160 £80

A large Chinese bronze baluster shaped vase, with short neck, 18in. high. $180 £90

One of a pair of 19th century cloisonne vases. $190 £95

One of a pair of Japanese bronze bottle shaped vases with figures of birds in high relief, 37cm. high. $230 £115

A 19th century Japanese cloisonne enamel bulbous shape bottle vase, 13in. high. $230 £115

Large 19th century Oriental embossed bronze bulbous vase surmounted by a Kylin. $240 £120

Japanese Art Nouveau Zeit Geist cloisonne vase, with damaged base, 46cm. high. $300 £150

One of a pair of late 18th century bronze Chinese vases, 17½in. tall. $360 £180

One of a pair of Chinese 12in. cloisonne enamel covered vases. $370 £185

One of a pair of 19th century ormolu mounted cloisonne enamel vases. $500 £250

119

VASES

A good 18th century Chinese cloisonne vase, 11½in. tall.
$760 £380

Early 17th century cloisonne vase, 23in. high.
$960 £480

A fine and large Chinese gold splash bronze vase, the base with Hsuan Te mark, 18¾in. high.
$1,080 £540

One of a pair of large 19th century vases, 24in. high.
$1,130 £565

Mid 17th century Chinese cloisonne pear shaped vase, 20½in. high.
$1,200 £600

Elaborate Shibayama vase and cover, mother-of-pearl flower-heads, 6in. high, late 19th century.
$1,440 £720

One of a pair of 19th century cloisonne vases. $1,560 £780

One of a pair of cloisonne vases, royal blue on powder blue, circa 1860. $1,500 £750

One of a pair of Japanese cloisonne enamel vases, 36in. high, circa 1900. $1,960 £980

Japanese bronze vase, signed by Dai Nihon Kyoto ju Bunryu. $2,000 £1,000

A fine Japanese cloisonne enamel vase, 3ft. high. $2,060 £1,030

One of a pair of Chinese cloisonne vases, 31in. high. $2,300 £1,150

VASES

A pair of cloisonne enamel beaker vases of the Ch'ien Lung period, 42cm. high. $2,640 £1,320

Large pair of Chinese cloisonne enamel vases, the blue ground decorated with a floral design. $3,080 £1,540

A large cloisonne enamel vase, 50in. high. $4,200 £2,100

One of a pair of gold and Shibayama vases signed Yaschika, 12½ins. high. $6,600 £3,300

A pair of Kintani inlaid bronze vases and stands, 116.2cm. high. $7,000 £3,500

A pair of Komai style bronze vases, 17½in. high. $7,200 £3,600

INDEX